PROGRESS IN

WORKBOOK 2

GEOGRAPHY

UNITS 6–10

KEY STAGE 3

JO COLES
ELEANOR HOPKINS
CATHERINE OWEN

T0187249

HODDER
EDUCATION
LEARN MORE

NAME: ....

CLASS: ....

The Publishers would like to thank the following for permission to reproduce copyright material.

Photo credits

p. 10 © CPQ - stock.adobe.com; **p. 11** © Catherine Owen; **p. 15 b** © Air Images Ltd; **p.17** © Marian Kamensky, www.CartoonStock.com; **p. 23** © WaterAid/Joey Lawrence; **p. 37 l** Image Landsat/Copernicus © 2018 Google. Data SIO, NOAA, U.S. Navy, NGA, GEBCO © 2009 GeoBasis-DE/BKG. Contains BGS Geology 625k Data © UKRI 2018, https://www.bgs.ac.uk/data/services/digmapgb625kml.html; **p. 37 r, p. 39** © David Gardner; **p. 40** © Jo Coles; **p. 42 l and r** Google Earth © Infoterra & Bluesky. Image NASA; **p. 44** © A.P.S.(UK)/Alamy Stock Photo; **p. 45** Sameer Burle, http://www.floodmap.net. Map data © 2019 Google **p. 49** © Valerii Shanin/Alamy Stock Photo; **p. 52** © Shutterstock/Pagespics.

Acknowledgements

All OS maps used throughout this book have been reproduced from Ordnance Survey mapping with permission of the Controller of HMSO. © Crown copyright and/or database right. All rights reserved. Licence number 10003470.

Every effort has been made to trace all copyright holders, but if any have been inadvertently overlooked, the Publishers will be pleased to make the necessary arrangements at the first opportunity.

Although every effort has been made to ensure that website addresses are correct at time of going to press, Hodder Education cannot be held responsible for the content of any website mentioned in this book. It is sometimes possible to find a relocated web page by typing in the address of the home page for a website in the URL window of your browser.

Hachette UK's policy is to use papers that are natural, renewable and recyclable products and made from wood grown in well-managed forests and other controlled sources. The logging and manufacturing processes are expected to conform to the environmental regulations of the country of origin.

Orders: please contact Hachette UK Distribution, Hely Hutchinson Centre, Milton Road, Didcot, Oxfordshire, OX11 7HH. Telephone: +44 (0)1235 827827. Email education@hachette.co.uk Lines are open from 9 a.m. to 5 p.m., Monday to Friday. You can also order through our website: www.hoddereducation.co.uk

ISBN: 9781510428065

© Eleanor Hopkins, Catherine Owen, Jo Coles 2019

First published in 2019 by

Hodder Education,

An Hachette UK Company

Carmelite House

50 Victoria Embankment

London EC4Y 0DZ

www.hoddereducation.co.uk

Impression number   10 9 8

Year   2023

All rights reserved. Apart from any use permitted under UK copyright law, no part of this publication may be reproduced or transmitted in any form or by any means, electronic or mechanical, including photocopying and recording, or held within any information storage and retrieval system, without permission in writing from the publisher or under licence from the Copyright Licensing Agency Limited. Further details of such licences (for reprographic reproduction) may be obtained from the Copyright Licensing Agency Limited, www.cla.co.uk

Cover photo © airquiplay 77-stock.adobe.com

Illustrations by Aptara Inc.

Typeset in India by Aptara Inc.

Printed in Great Britain by Ashford Colour Press Ltd.

A catalogue record for this title is available from the British Library.

MIX
Paper from responsible sources
FSC™ C104740
FSC
www.fsc.org

Contents

Contents

COMPLETED

© Hodder Education 2019

COMPLETED

10 How is Asia being transformed?

The uses of the River Nile

The River Nile in Africa is one of many important rivers around the world. At 6,695 km long it is the longest river in the world. Here are some facts that show the reasons why the River Nile was important to people in Ancient Egypt and Modern Egypt:

How the River Nile was used in Ancient Egypt

Rain was unreliable in Ancient Egypt, so the Egyptians relied on the Nile for water, transportation and good soil for growing crops.

Papyrus reeds from the river were used to make paper and boats; spears and nets were used to catch fish.

How the River Nile is used in Modern Egypt

The Aswan Dam was built in the 1960s to generate hydroelectricity and to control floods to make them less dangerous.

The river is still important for irrigation (watering crops) and tourists enjoy Nile River cruises.

Complete the mind map below using this information about the River Nile or carry out research into another river so that you can complete the mind map. You could investigate the Ganges, the Amazon or a river of your choice.

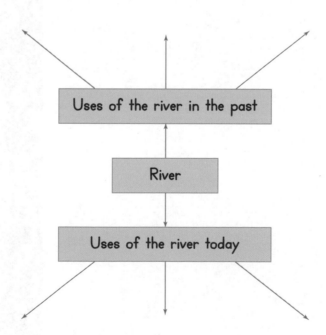

© Hodder Education 2019

Student's Book
pages 104–105

The water cycle – key terms

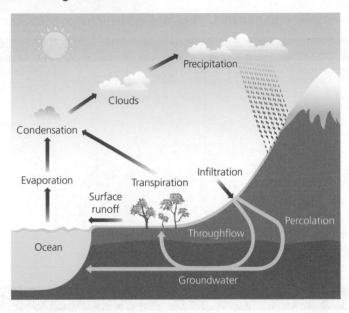

The water cycle

Using images alongside words helps you to remember important words and facts.

1 Complete the table below by using words and drawings to show what the terms mean.

Evaporation	The sun heats water and it changes into water vapour.	
Condensation		
Precipitation		
Infiltration		
Percolation		
Throughflow		
Groundwater		
Surface run off		
Transpiration		
Water cycle		

River processes

1 Add diagrams to the boxes below to turn this sheet into a poster for river processes.

Erosion

Hydraulic action	Abrasion	Attrition	Corrosion

Transportation

Traction	Saltation	Suspension	Solution

Deposition

2 What conditions are needed for the river to deposit its load?

_____

_____

_____

© Hodder Education 2019

Profiles of a river

1 Complete the paragraph using the words in the box.

v-shaped	long	steep	vertically

The _____ profile shows changes in the gradient of a river as it flows from its source

to its mouth. In the upper course of a river the gradient is usually _____ and the river

erodes its channel _____ This creates _____ valleys.

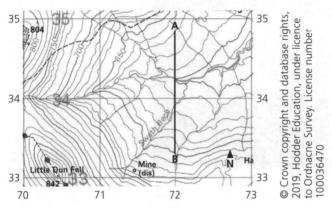

Map of the River Tees, scale 1:50 000

© Crown copyright and database rights, 2019, Hodder Education, under licence to Ordnance Survey. License number 100036470

2 The cross profile of a river is a slice taken across a river or its valley. Use the line from A–B on the map of the River Tees above to draw a sketch of its cross section.

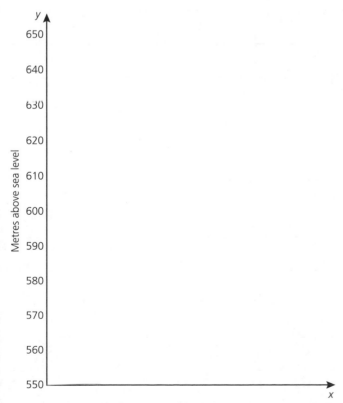

3 Does this cross section show the river in its upper, middle or lower course? Give evidence to support your answer.

_____

_____

_____

_____

_____

_____

_____

_____

_____

_____

_____

Waterfalls

Waterfalls are not only interesting to geographers because of how they form and change over time. They can also be extremely beautiful, as this photograph of Niagara Falls shows.

Niagara Falls

1 Imagine that you are a travel writer visiting Niagara Falls. How would you describe the waterfall? Include the following terms in your description: plunge pool, erosion, undercutting, gorge.

_____

_____

_____

_____

_____

_____

_____

2 How will this waterfall change over time?

_____

_____

_____

_____

_____

_____

© Hodder Education 2019

Testing the Bradshaw model

Upstream   **Downstream**

Channel depth

Occupied channel width

Mean velocity

Discharge

Volume of load

Load particle size

Channel bed roughness

Gradient

The Bradshaw model

In your lesson you have looked at how to measure the width, depth and speed of a river, using the example of the River Holford.

This photograph shows Megan measuring the long axis (longest length) of a pebble she picked up at one of the sites she studied along the River Holford.

1 Which aspect of the Bradshaw model is Megan investigating using this method?

_____

_____

_____

2 How many pebbles do you think she should measure at each site? Explain why you think this is a good sample size to use.

_____

_____

_____

3 Megan tried to select pebbles randomly by closing her eyes before picking them up. Why do you think she wanted to choose pebbles randomly?

_____

_____

_____

4 Can you suggest improvements to the method Megan is using?

_____

_____

_____

A local river fact file

Carry out your own research into a river near your home or school. Complete the fact file below.

Name of river:

_____

Location of river:

_____

Description of river:

_____

_____

Which settlements are located near to this river?

_____

_____

_____

Why are these settlements located near to this river?

_____

_____

_____

How is this river used for leisure and recreation?

_____

_____

_____

Do any industries use this river? If so, how do they use it?

_____

_____

_____

How important is this river to you? Explain your answer.

_____

_____

_____

© Hodder Education 2019

Damage caused by flooding

1 Enter your postcode at https://flood-map-for-planning.service.gov.uk/. Is your home at risk of flooding?

_____

_____

2 Suggest reasons for the level of flood risk. Consider how close you are to rivers, local relief (flat/hilly), height of land and climate.

_____

_____

_____

_____

> Flood damage, such as that caused by the 2015 floods in York, can be very upsetting for the people affected.
>
> We can divide the damage caused as tangible or intangible. It is possible to put a price on tangible damage and to repair/replace it. However, intangible damage means that something that is irreplaceable has been damaged, such as an old photograph or a special letter.

3 If your home were to flood to a depth of 30 cm, what tangible damage would happen?

_____

_____

_____

4 Would the flood cause any intangible damage? If so, what would the damage be?

_____

_____

_____

5 What could you do to minimise the chance of your home being flooded?

_____

_____

_____

_____

6.9 How can flooding be managed?

Soft-engineering

The long-term vision for managing flooding of the River Ouse includes ideas for managing the drainage basin to reduce the amount of water reaching the river.

- The Environment Agency will work with farmers to manage run off.
- Washlands and reservoirs will mean that water spreads naturally across farmland, reducing the rate at which it arrives in York.
- Tree planting can slow water down or prevent it reaching rivers.

This is an example of soft engineering – working with nature to reduce the flood risk.

Think back to your work on the water cycle at the start of this unit to help you answer the following questions.

1 A washland is an area of land which is allowed to flood in the upper course of the river. Why will floods in this area cause less damage than floods further downstream?

_____

_____

_____

2 How will the way farmers use their land influence the amounts of run off, infiltration and transpiration?

_____

_____

_____

3 Draw a diagram to show how trees intercept precipitation, use water and hold soils together.

4 Research a flood defence scheme in your local area. What methods are used? Why?

_____

_____

_____

_____

© Hodder Education 2019

Student's Book pages 120–121

Rivers review

Complete this sheet to help you organise your knowledge for this unit.

What are rivers and how does water get into them?

Rivers are…

Water gets into them from…

How do weathering, erosion and transportation create river landforms?

Waterfalls are formed by…

Meanders are formed by…

What do river landforms look like on an OS map?

This map extract shows a confluence at…

It shows a meander at…

Why are rivers important to people?

Water supply:

Settlement:

Industry:

© Crown copyright and database rights, 2019, Hodder Education, under licence to Ordnacne Survey. License number 100036470

7.1 What is development?

Families around the world

1 Fill in the tables below to create a comparison of families at different levels of development around the world. You will need to carry out research on the website Dollar Street to find a family and look around their home. Select a photograph you think best shows their quality of life.

Use this website to investigate contrasting families around the world: www.gapminder.org/dollar-street

Add a photograph of a family found from Dollar Street.

Add a photograph of another family at a contrasting level of development found from Dollar Street.

Name of family	
Country	
Income	
How many family members are there?	
What is their house like? Describe the areas for cooking, cleaning and sleeping.	
What may this show about their quality of life? Remember all aspects of development shown in the DCR – Social, Economic, Environmental, Political.	
Interesting facts	

Name of family	
Country	
Income	
How many family members are there?	
What is their house like? Describe the areas for cooking, cleaning and sleeping.	
What may this show about their quality of life? Remember all aspects of development shown in the DCR – Social, Economic, Environmental, Political.	
Interesting facts	

2 Write a paragraph to describe the similarities and differences of the families you have chosen.

_____

_____

_____

_____

_____

 © Hodder Education 2019

7.2 How is money spread around the world?

Analysing a cartoon

Below is a cartoon showing an artist's view of how wealth is spread around the world. Study it carefully and answer the questions below. Use a map of the global distribution of GNI per capita to support your findings.

How wealth is found around the world

1 Write a list of countries which may be found either side of the seesaw.

   Left: _____

   Right: _____

2 What is the artist suggesting about how wealth is spread around the world?

   _____

   _____

   _____

3 Map C in Lesson 7.2 showed countries split into four categories of income. Write a critique of this cartoon. Which countries has the artist missed out?

   _____

   _____

   _____

   _____

   _____

Comparing GNI per capita and life expectancy

Complete the scatter graph showing the relationship between wealth and health for the countries shown in Table A. On the x-axis plot the GNI per capita, on the y-axis plot the life expectancy. Remember to give your graph a title and label the axis. Below the graph, first describe your findings, then suggest reasons for the link.

	GNI per capita	Life expectancy
China	8260	76.0
Malawi	320	63.9
Nepal	730	70.0
Norway	82,330	81.1
Sierra Leone	490	51.3
UK	42,390	80.8
USA	56,180	79.2

Table A The GNI per capita and life expectancy rankings for a selection of countries at different levels of development, 2015

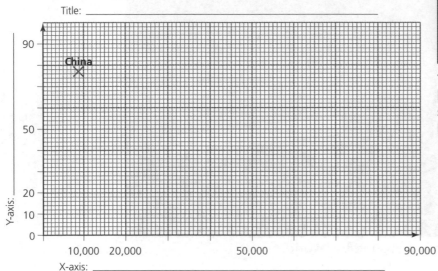

Title: _____

X-axis: _____

Descriptions of findings

_____

_____

_____

_____

Description tips
- First **describe** the general pattern.
- Use **evidence** of country names and figures of development.
- Suggest any **anomalies**.

Explanation of findings

_____

_____

_____

_____

Key terms
- Development
- Economic
- Health
- Social
- Quality of life

© Hodder Education 2019

7.4 How can development change over time?

Inventions fact file

Research one of the inventions from the list below and complete a fact file to show your findings.

> Light bulb – Diesel engine – Fertiliser – Antibiotics – Container

Photograph or sketch of invention

Key facts about the inventor (date born, country of birth)	Key facts about the invention (date, use)

What was life like before the invention?	How has the invention shaped the way we live?

How did the invention help the development of countries and people?

7.5 What is the global development map missing?

Dollar Street

Revisit the Dollar Street website and find evidence of inequality within one country.

Country _____

Name of family	
Income	
Describe their quality of life	

Name of family	
Income	
Describe their quality of life	

Describe how inequality occurs within the country you have chosen.

_____

_____

_____

_____

_____

_____

_____

_____

Name of family	
Income	
Describe their quality of life	

STRETCH: Print a photograph from each family and annotate the images to show the differences in quality of life.

© Hodder Education 2019

Earthquake in Nepal

In your lesson you investigated the causes of uneven development. Read the article about Nepal and answer the questions below.

One million people fall below poverty line in Nepal after earthquake

June 14, 2015

On 25 April 2015, a 7.9 magnitude earthquake occurred in Nepal killing 8,800 people, injuring thousands and leaving 1 million homeless.

According to the Human Development Report (2014), 23.8 per cent of people in Nepal were already living in poverty before the disaster. However, this earthquake has pushed even more people below the poverty line. Homes have been destroyed and countless possessions have been lost by many, from necessities like basic kitchen utensils to valuable items such as jewellery.

The earthquake has severely affected agriculture and people will become poor because they have lost precious resources such as seeds and livestock. People are dependent on these to earn income as farmers.

Education facilities have been destroyed and children cannot go to school. Water is contaminated and there is a lack of clean sanitation facilities. This will impact the health of both adults and children. These are all crucial elements of human development.

Over the last three years, Nepal's Human Development Index has been steadily increasing. However, the earthquake has impacted education, healthcare and the economy. This shows just how vulnerable the population in Nepal is.

> The **Human Development Index (HDI)** is a method of measuring development in which income, life expectancy and education are combined to give an overview.

1 How many people were killed in Nepal from the earthquake?

_____

2 Apart from deaths and injuries, how did the earthquake impact Nepal?

_____

_____

3 Nepal suffers from many earthquakes. Why will earthquakes like this cause challenges for the development of Nepal?

_____

_____

_____

4 STRETCH: What will Nepal need to do to improve the HDI for the people living there? Why?

_____

_____

_____

_____

The Magic Washing Machine

Carry out an internet search for 'Hans Rosling The Magic Washing Machine TED' and click on the video. As you watch the video, answer the following questions.

1 How did Hans' grandmother wash clothes? Underline the correct answer.

   a) Heating water over firewood and washing

   b) Going to a lake

   c) In a bath

2 What was purchased by the family which made everyone celebrate? Underline the correct answer.

   a) An iron

   b) A washing machine

   c) A tumble dryer

3 Name three different machines people in Sweden and wealthy countries use in the home.

   1 _____

   2 _____

   3 _____

4 How do people who live below the poverty line cook their food and wash clothes?

   _____

5 How many billion people have a washing machine? _____

6 Why does everyone want a washing machine? _____

   _____

7 What would be a negative if everyone had a washing machine?_____

   _____

8 Why may literacy rates (the number of people who can read and write) and HDI (Human Development Index) improve if women spend less time doing housework?

   _____

   _____

9 Are there any other inventions that have revolutionised the lives of women?_____

   _____

10 What are the implications on gender equality when a country industrialises?

   _____

   _____

© Hodder Education 2019

Social media for global campaigns

WaterAid is an NGO (non-governmental organisation) like ActionAid. It creates posters for social media to encourage people to donate money. This money could be used to support projects which improve lives for people living in poverty.

In the example here, it shows that WaterAid works with UK aid. UK aid is led by DFID (Department for International Development) to end extreme poverty. Here, if people donate, the government will double the donation to WaterAid.

1   Who is the poster created by?

_____

_____

2   How is UK aid supporting the campaign?

_____

_____

_____

_____

_____

_____

WaterAid

3   What are the benefits of using social media to raise awareness?

_____

_____

_____

4   Go online and look at the social media platforms (Twitter, Instagram) of non-governmental organisations such as WaterAid, Oxfam, Save the Children. You could also carry out a search in Google. On a blank sheet of paper, create a collage of the best campaigns you find.

5   Write a list of what makes a successful campaign.

_____

_____

_____

_____

7.9 What are Sustainable Development Goals?

The Sustainable Development Goals comic strip

Create a cartoon strip about one of the following Sustainable Development Goals, showing how it is important for the future of the planet and its people. Follow this link http://www.comicsunitingnations.org/ to see how comics have been used by the United Nations to explain the goals.

1 No poverty
2 Zero hunger
3 Good health and well-being
4 Quality education
5 Gender equality
6 Clean water and sanitation
8 Decent work and economic growth
10 Reduced inequality
16 Peace and justice strong institutions

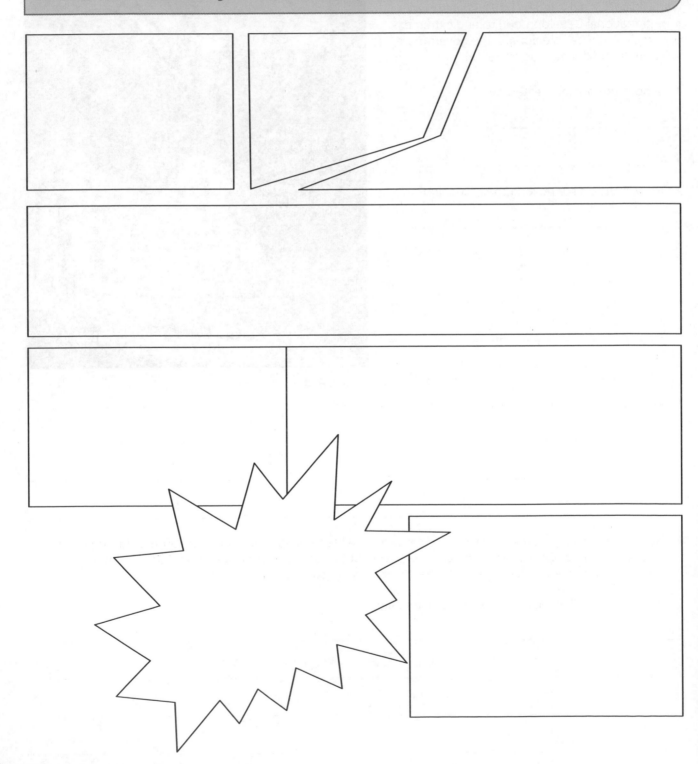

© Hodder Education 2019

Concept mapping

Key terms

Aid	BRICS	Climate	Development
Education levels	Ecological footprint	Gender equality	GNI
HDI	Inequality	Inventions	Life expectancy
NGOs	Poverty	Quality of life	Sustainable Development Goals

1   Copy out each of the key terms into the space below, making sure there is a lot of space around each term.

2   Under each key term, write a definition.

3   Draw a line between two or more terms that are linked together. Explain the link you have made. Continue this to link more terms and create a concept map. This will help you make links between the concepts and terms in Unit 7.

Predicting population growth

According to the UN, half of world population growth from now until 2050 is forecast to be in India, Nigeria, Congo, Pakistan, Ethiopia, Tanzania, the USA, Uganda and Indonesia.

1 Mark the nine countries listed above on the map below.

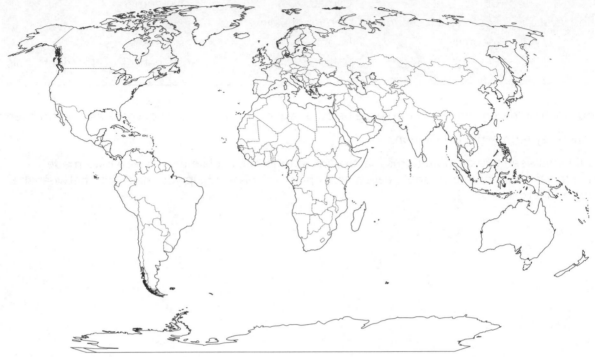

2 Describe the distribution of the countries shown on the map.

_____

_____

_____

_____

_____

3 Do you think it is possible for the UN to forecast population growth up until 2050? Give reasons to support your answer.

_____

_____

_____

_____

_____

_____

© Hodder Education 2019

8.2 Where does everyone live, and why?

Population density

1 Define the following terms:

Population distribution _____

Densely populated _____

Sparsely populated _____

2 Complete the table below.

Place	Sparsely or densely populated?	Possible reasons for population density
The Himalayas in Nepal		
Los Angeles in the USA		
The Sahara Desert in North Africa		
The South East of the UK		

3 Do you live in an area which is densely or sparsely populated? Why do you think this is?

_____

_____

_____

_____

8.3 How can we describe the structure of a population?

The Demographic Transition Model

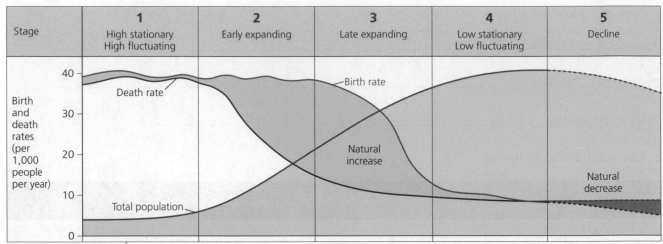

Stage	1 High stationary High fluctuating	2 Early expanding	3 Late expanding	4 Low stationary Low fluctuating	5 Decline

Stage 1: Birth rate and death rate are high but fluctuate due to disease, famine and war.
The total population stays low.
E.g. UK before 1780, Niger today

Stage 2: Death rate decreases due to improvements in medical care, scientific discoveries, hospitals and improved sanitation and water supply, but birth rate stays high.
The population grows as the natural increase becomes higher.
E.g. UK 1780–1880, Afghanistan today

Stage 3: Birth rate falls rapidly as the country develops. Use of contraception, due to government incentives, the changing role of women in society, or increased desire for material possessions rather than large families.
E.g. UK 1880–1940, Brazil today

Stage 4: Birth rate and death rate are low but fluctuate; steady, small natural increase.
E.g. UK today

Stage 5: Birth rate very low, falling below the death rate, leading to a natural decrease. Death rate increases slightly due to an ageing population.
E.g. Russia today

The Demographic Transition Model

Choose one of the stages of the Demographic Transition Model. If you were to visit a country at this stage in the model, what would it be like? Include the terms in the box below in your answer.

> Birth rate   Death rate   Population growth   Health   Access to contraception
> Role of women in society   Quality of life   Incentives

_____

_____

_____

_____

_____

_____

_____

_____

© Hodder Education 2019

Promoting population growth

The Chinese government introduced a one-child policy in 1970 and promoted this using a poster campaign.

In 2007 the Russian government introduced a programme to encourage population growth. Payments of £11,000 were introduced for mothers with more than one child. The money can be put towards buying a house or towards the child's education, or be deposited in the mother's pension scheme.

Design a poster in the box below which could be used to promote population growth in Russia.

Migration

Migration doesn't just mean moving from one country to another, it could mean moving from the countryside to a city or from one city to another.

1 Define the key terms below:

Immigrant _____

_____

Voluntary migrant _____

_____

Forced migrant _____

_____

Refugee _____

_____

2 Think about a place that you would like to migrate to when you are older. Where would you go?

_____

3 Explain why you would like to move to this place. Include specific details about the place and use key terms such as push factors, pull factors and intervening obstacles.

_____

_____

_____

_____

_____

_____

_____

_____

_____

© Hodder Education 2019

Applying Lee's migration model to migration from Mexico to the USA

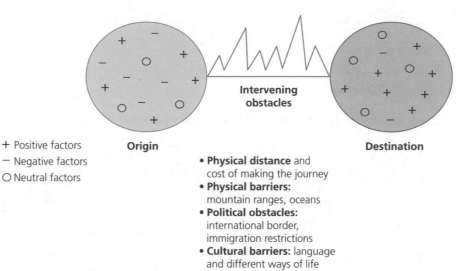

+ Positive factors
− Negative factors
○ Neutral factors

Origin

Destination

Intervening obstacles

- **Physical distance** and cost of making the journey
- **Physical barriers:** mountain ranges, oceans
- **Political obstacles:** international border, immigration restrictions
- **Cultural barriers:** language and different ways of life

E.S. Lee's migration model

In 2007, 12.6 million Mexican-born immigrants were living in the USA. Produce your own version of E.S. Lee's migration model and what you learnt in the lesson to show:

- the push and pull factors that persuade people to move from Mexico to the USA
- examples of intervening obstacles they may face.

8.7 What is urbanisation?

How Mumbai's population is changing

Year	Population (millions)
1990	12.4
2015	21.0
2030	27.8

Population of Mumbai, 1990–2030

1   Draw a line graph on the axes below to show how Mumbai's population will change from 1990–2030.

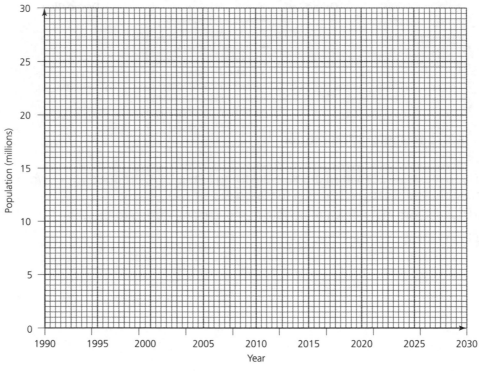

2   How many times bigger is the population of Mumbai predicted to be in 2030 than in 1990?

_____

3   Rural–urban migration is one of the main reasons for the growth of Mumbai. What does rural–urban migration mean?

_____

_____

_____

4   Suggest one reason for rural–urban migration in India.

_____

_____

_____

   © Hodder Education 2019

8.8 How did urbanisation change Southampton? Part 1

8.8

Student's Book
pages 156-157

Moving to Southampton in 1840

Southampton in the 1840s was growing rapidly as people moved to work in the new docks and industries such as grain milling and tobacco processing. Houses were built for these workers, but there were no bathrooms, running water or toilets, so disease was a massive problem.

Write an account or poem telling the story of someone moving to Southampton to work in 1840.

Think about:
- reasons for leaving the countryside (push factors)
- reasons for moving to Southampton (pull factors)
- obstacles that they had to overcome to get to the city
- what life was like once they reached the city.

_____

_____

_____

_____

_____

_____

_____

_____

_____

_____

_____

_____

_____

_____

8.9

Student's Book
pages 158–159

8.9 How did urbanisation change Southampton? Part 2

The Burgess land-use model

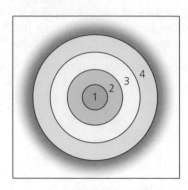

The Burgess land-use model

1 What does the Burgess land-use model tell us about how some cities grow?

_____

_____

_____

_____

2 Does Southampton fit the Burgess land-use model?

_____

_____

_____

_____

3 Choose a town or city in your local area. Does it fit the Burgess model?

Name of town/city: _____

Ways in which it fits the Burgess model: _____

_____

_____

Ways in which it doesn't fit the Burgess model: _____

_____

_____

© Hodder Education 2019

8.10 One planet, many people: how are populations changing? Review

8.10

Population and urbanisation – key terms

Complete the table below to show the definitions of key terms from this unit.

Key term	Definition
Birth rate	
Census	
Death rate	
Demographic Transition Model	
Natural decrease	
Natural increase	
Over-populated	
Population	
Population distribution	
Population pyramids	
Under-populated	

Field sketch

A field sketch is a drawing that geographers make. It is not a piece of art, because a field sketch has descriptive labels (annotations) to describe the landscape in more detail. The annotations should include:

- the direction you are facing
- the name of the location
- physical and human features
- what you can see and hear
- even what you might smell or touch or taste!

It should give the viewer a really good idea of what this place is like.

Think about a coastal location you have been to, or one you would like to go to (you can look on the internet to help you). Draw a detailed field sketch in the box below. Remember those descriptive labels! Think about how the location is used by people (social, economic and environmental uses). You could use different coloured pens for the annotations to make them clearer. Draw in pencil. Use a ruler for straight lines and buildings.

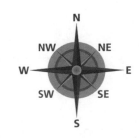

© Hodder Education 2019

Looking at rock types

A geological map of the UK

Flamborough Head

This is a geological map from Google Earth and the British Geological Survey (BGS). It is created by geologists and used to show rock and soil type.

Geologists study rock types and how they behave. Harder rocks (like chalk and granite) tend to be stronger and resist erosion for longer, whereas softer rocks (like clay and mudstone) tend to erode more rapidly. Our rocks are formed in layers, with deep bedrock being different to the surface rocks. For example, the Holderness coast bedrock is chalk but on the surface is glacial till – soft material moved from far away and left there by glaciers.

1 Look at the image above. What type of rock is at Flamborough Head?

_____

Go to the British Geological Survey website at http://mapapps.bgs.ac.uk/geologyofbritain/home.html.

2 What is the rock type where you live?

_____

3 Is this rock type hard or soft? How do you know?

_____

_____

4 Use the British Geological Survey website to investigate the type of rock at a different coastal location of the UK (not Flamborough). How do places that have clay or mudstone compare to places that have granite?

_____

_____

_____

_____

9.3 What forms of erosion take place on the coast?

Erosion: extended writing

Being able to 'speak like a geographer' is very important. It means you try to use key words to describe and explain your answers. So far you have learned about geology, weathering and types of erosion. Now is your chance to put this into some extended writing.

Write a short story, about 250 words, to describe how a very small, very rounded clay pebble once started its life as part of a very big cliff at Flamborough Head. Try to include what happened to it to make it leave the cliff, and why it has changed shape to become round. Be creative, but still try to use key words.

_____

_____

_____

_____

_____

_____

_____

_____

_____

_____

_____

_____

_____

_____

_____

_____

_____

_____

_____

© Hodder Education 2019

Headlands and bays

1 Annotate the photograph of Selwicks Bay at Flamborough Head, to show:

   ● geology type and how this influences the strength of the rocks

   ● what types of landforms are visible

   ● evidence of erosion and weathering.

2 Use the internet to research a different example of a headland or a bay coastal landscape. In the space below, glue in a selection of photographs that show what the landscape is like.

3 Now write a description to outline how the landscape has been eroded over time. Think about what it may have looked like thousands of years ago, and how it might change in future.

_____

_____

_____

_____

_____

9.5 How does transportation change the coastline?

Analysing a photograph

1 Label the photograph to show the swash and, the backwash and where the waves break.

2 Describe the waves that are approaching the beach. (Think about the height of the wave, whether they look steep, whether they look like the swash or the backwash is strong, etc.)

_____

_____

_____

3 Are these waves constructive, or destructive? What is your proof?

_____

_____

4 How will these waves affect the shape of the beach over time? What problems could be caused?

_____

_____

_____

_____

5 How might longshore drift or waves affect different uses of the beach by different groups of people over time?

_____

_____

_____

© Hodder Education 2019

Deposition landform case study

Choose an example of a deposition landform:

> tombolo – beach – bar – spit – salt marsh

Use the internet to find an example of this landform from a real place somewhere in the world. Use this page to create a case study, using the checklist on the right for guidance.

- Map of location
- Description of location
- How the landform was created
- How it is used by people
- Whether this landform is at any risk of erosion

9.7 How has life on the Holderness coast changed?

Changes on the Holderness coast

2003

2017

Study the two photographs of the Holderness coast near Cowden.

1   Make a suggestion: what do you think the buildings are for? How is the land here used by people?

_____

_____

2   How has the shape of the coastline changed from 2003 to 2017?

_____

_____

3   If you lived at Cowden, how would you be feeling? What emotions would you have and why? What worries would you have for the future?

_____

_____

_____

_____

4   If you were a member of the Environment Agency, or the local council, what would you propose doing for this area? How could you help?

_____

_____

_____

_____

_____

© Hodder Education 2019

Different coastal management strategies

1 Use what you have learned about the different types of sea defences – sea walls, groynes, gabions, rock armour, beach recharge – and choose **ONE** of the following activities to demonstrate what you have learned so far.

a) If you live near a beach and can safely visit it, collect your own photographs/sketches of the different sea defences and create an annotated collage. Label what the sea defences are and why they are used.

b) If you do not live near a beach, you could investigate one location using the internet and create an annotated collage of printed photographs. Label what the sea defences are and why they are used.

c) Carry out a survey in class or at home on the value of sea defences. Make a note of the different opinions. What sorts of sea defences are more popular? Which are least liked? Why might this be the case?

d) Imagine you are living in a town that suffers from coastal erosion like Mappleton. Write a letter to the council to tell them why you need sea defences, which ones you think would help most, and why they will help protect your home.

2 In your opinion, which sea defence strategy is best? Why? Write a piece of persuasive writing to convince a town council to choose the strategy you think is best.

Investigating Spurn Head

This is Spurn Head, the spit at the southern end of the Holderness Coast. It is also a Site of Special Scientific Interest protected from erosion.

1 Look at the aerial photograph of Spurn Head. Annotate the photograph to show how this spit has formed.

2 Annotate on the photograph where and why you think this spit is at risk of being eroded.

3 Spurn Head spit has changed positions many times over hundreds of years. It is at risk of being breached by the sea. Explain what causes this.

_____

_____

_____

_____

Spurn Head

4 Research Spurn Head. Find out why it is defended and protected from erosion.

_____

_____

_____

_____

5 In your opinion, should Spurn Head be protected? Consider the costs and benefits of sea defences like groynes. Are they worth it to protect Spurn Head? You could create a table to compare the advantages and disadvantages to help organise your thoughts. Remember, you should form a conclusion based on fact – so consider the financial cost, and how successful the defences are, and whether the land is valuable enough.

_____

_____

_____

_____

9.10 What happens where the land meets the sea? Review

9.10

Student's Book
pages 180-181

What will happen if sea levels rise?

Study the maps from http://www.floodmap.net

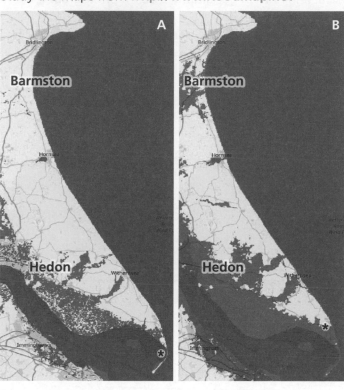

The * shows where Spurn Head spit is located.
They show the predicted impact of:

A 2m sea level rise

B 7m sea level rise along Holderness.

1 Why do you think sea levels might rise in the future? Suggest what could cause this.

_____

2 What is predicted could happen to Spurn Head in future if the sea level rises?

_____

3 What happens to the rest of the Holderness area inland, such as along the river estuary by Hedon or at Barmston?

_____

_____

4 Do you think that sea level rise would have an impact on how severe erosion is? Why?

_____

_____

5 What might happen to vulnerable coastal places like Flamborough Head sea stacks?

_____

_____

_____

6 What might happen to the sea defences that are in place along the Holderness coast? Will they be effective if sea levels rise? Explain your answer.

_____

_____

_____

7 Suggest whether the Holderness coast might require other sea defences in future, and why this might be the case.

_____

_____

10.1 Diverse and dynamic: how is Asia being transformed?

Exploring Asia

1 Complete the research to show Asia's diverse physical and human geography. Insert a photograph or draw a sketch of each feature. The first two ideas are provided for you.

Asia's largest river:
_____

Asia's highest mountain:
_____

2 Describe why Asia is a diverse continent.

_____

_____

_____

_____

© Hodder Education 2019

Monsoons

Type in the following web link into your internet browser, www.metoffice.gov.uk/videos/5502225157001 or search for 'Met Office. What is a monsoon? Video' in a search engine. Once you have watched the video, answer the following questions.

1 The word '**monsoon**' comes from the **Arabic** word *mausim*. What does this word mean? _____
_____

2 What does the dry season change into during the monsoon?_____
_____

3 What percentage of the world lives in a monsoon climate?_____

4 What percentage of annual rainfall do many parts of India receive?_____

5 What powers the monsoon?_____
_____

6 What heats up more quickly: land or water?_____

7 What temperature can land in India be in May?_____

8 How much cooler are the seas?_____

9 Why does the cool air move inland?_____
_____

10 What direction is the monsoon?_____
_____

11 What geographical feature stops the clouds, forcing them to rise and rain?_____
_____

12 Why is the monsoon essential for communities in Asia?_____
_____

13 What issues arise if the rains are too heavy?_____
_____

14 Why does the monsoon end?_____
_____

Decision-making exercise: Floods in Bangladesh

Situation: You are in charge of the Humanitarian Coordination Task Team (HCTT) for Bangladesh. This means you have been given money raised to help communities who have been affected by the floods. You have US$6 million. 330,000 people are affected, half of whom are children.

Task: Read through the options carefully. Complete the list in the table at the bottom of the page to show what things you will spend the money on to help the community. Remember you only have $6 million.

Child Protection	Nutrition	Early Recovery
Building shelters for children whose parents have died in the floods and provide a safe space for children in the worst affected areas. ($400,000)	Provide emergency food, health and medical services to prevent children under five becoming ill ($500,000)	Organising a 'cash for work scheme' where farmers who have lost their job will be paid to rebuild the community. ($1,800,000)
Education	**Water Sanitation And Hygiene**	**Food Security**
Rebuilding schools so children can go to school. ($1,000,000)	Ensure safe drinking water and improved sanitation (toilet and sewage) facilities to the affected population. ($2,000,000)	Donating food to the 30,000 most vulnerable households for three months. ($3,500,000)
Gender Equality		**Shelter**
Support pregnant women that may become ill due to a lack of access to safe water. ($2,000,000)		Provide emergency places to live and sleep to the people whose homes were destroyed during the floods. ($2,500,000)

Spending Decision	Cost	Reason for choice
	Total:	

 © Hodder Education 2019

The impacts of deforestation

Look at the photograph below showing the impacts of deforestation.

1. What is the correct definition for deforestation?

  a. The cutting down of trees

  b. The planting of trees

  c. Creating a forest

2. Which biome is under threat from deforestation?

  a. Grassland

  b. Desert

  c. Mountain

3. Where in Nepal does deforestation occur?

  a. The Amazon

  b. The Himalayas

  c. The Ganges

4. Why does deforestation occur?

  a. For fuelwood

  b. To grow crops

  c. For timber

5. Draw a sketch of the photograph below showing the impacts of deforestation. Annotate it clearly to explain the impacts. Use the information in the box to help you with your annotations.

The impact of deforestation

> **Can you annotate your sketch with the following points?**
> - Soil compaction
> - Water run-off
> - Erosion
> - Lack of evaporation and transpiration
> - Habitats destroyed

STRETCH: On a separate piece of paper, write a letter to the government of Nepal appealing to them to stop deforestation. Use convincing language to describe the problems caused to the environment and why the biome should be protected.

10.5 Why is the population of Asia diverse and dynamic?

Population policies

In your lesson you looked at Japan's population policies. Japan has an ageing population. One issue is that there are not enough people of working age (16-64 years old) to work in jobs to support the whole population and the economy. The Japanese government has two main population policies to help solve this. Create a poster on **one** of the following policies below.

Policy 1: The Japanese government want to encourage couples to have more children. Create a poster directed at families. This policy will aim to increase the total population and number of workers. Use your knowledge of pro-natalist policies from chapter 8 to support you.

Policy 2: The Japanese government want to encourage people from other countries to move to Japan. Create a poster directed at people who live outside Japan to encourage them to migrate to Japan. This policy will aim to increase the number of people of working age who can earn money for the country. You can use your knowledge of push and pull factors to add to your poster.

Posters are most successful when they have bold colours and a clear message. Consider this in your initial plan.

© Hodder Education 2019

10.6 How is urbanisation changing lives in Karnataka, India? Part 1

10.6

Student's Book
pages 192–193

Diary of a migrant

Write a diary entry for a migrant who has moved. Give push and pull reasons and describe how it was difficult to leave home.

_____

_____

_____

_____

_____

_____

_____

_____

_____

_____

_____

_____

_____

_____

_____

_____

_____

_____

_____

_____

_____

_____

10.7

Student's Book
pages 194–195

10.7 How is urbanisation changing lives in Karnataka, India? Part 2

Field sketch skills

Geographers use field sketches to understand the landscape. Draw a sketch of the photograph below and annotate the human and physical features.

Hints

- Look at each section of the image so that you fit everything in. Consider what is in the foreground (front) of the image and the back.
- Leave out details of minor features, simple lines are best.
- Use a ruler to draw clean lines to the outside of the image to annotate your sketch.

China in the news

China is often in the news for technological advances and the growth of the economy. Below are some of the news article headlines which show China's rise.

- China is adding solar power at a record pace
- China's rise to global economic superpower
- China wants to build a $150 billion artificial intelligence industry
- Chinese investment into Africa soars in 2016
- China to overtake US as largest air travel market

Your task is to research why China is in the news today. Complete the table below to show your findings.

Title	
Newspaper/ Website	
What does the article say? Summarise it in no more than 50 words.	
How has this article made you think about China differently?	
Is the article trustworthy? Why?	

10.9 How is Asia developing into the most important global economic region?

Where is it made?

1 At home, look for the items in the table on the right and add the location of where they are made.

2 Plot all the locations on the world map below.

3 Connect the UK with these locations using a line. If more than one item comes from the same place, make the line thicker so that you can see where most items come from.

4 Describe where most of these items come from. How connected are you to Asia?

_____

_____

_____

_____

_____

_____

Item	Made in. . .
School top	
School shoes	
Non-school item of clothing	
Trainers	
Toothbrush	
Mobile phone	
Bed linen	
Kettle or toaster	
A mug	
Rucksack	
Extra:	

World map

© Hodder Education 2019

10.10 Diverse and dynamic: how is Asia being transformed? Review

10.10

Student's Book
pages 200–201

Mapping Asia

Throughout this unit you have been introduced to many maps. You will now create your own sketch map which will represent the diversity of Asia. Complete the map below by shading sections that illustrate Asia's diversity. You can either use a key (like the one below) or alternatively you can draw images to represent each area.

Consider both physical features (climate, mountains, rivers) and human features (densely populated areas, wealthy). In this map you are not limited to country areas or exact physical features. Be creative to show your understanding.

Key				
Mountainous		Flat		
Wet		Big city		
Densely populated		Sparsely populated		

Example key

Asia

PROGRESS IN

WORKBOOK 2

GEOGRAPHY

UNITS 6–10   KEY STAGE 3

Review and reinforce the skills, knowledge and understanding that you are developing throughout your Progress in Geography: Key Stage 3 course.

This Workbook accompanies your Student Book, providing extra support as you continue on your journey to become a good geographer.

- A range of activities focus on skills, knowledge and understanding
- Ideal for homework, classwork and independent study
- One Workbook page for every lesson in the Student Book

Also available:

Workbook 1: Units 1–5 (Single copy)
ISBN: 9781510428072

Workbook 1: Units 1–5 (Pack of 10)
ISBN: 9781510442993

Workbook 2: Units 6–10 (Pack of 10)
ISBN: 9781510443006

Workbook 3: Units 11–15 (Single copy)
ISBN: 9781510442986

Workbook 3: Units 11–15 (Pack of 10)
ISBN: 9781510443013

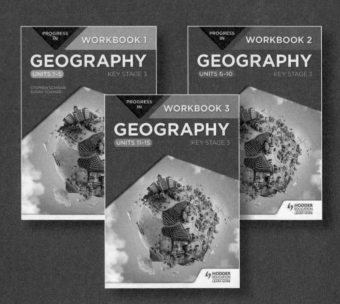

With special thanks to

The world's trusted geospatial partner

HODDER EDUCATION
t: 01235 827827
e: education@hachette.co.uk
w: hoddereducation.co.uk

ISBN 978-1-5104-2806-5

MIX
Paper from
responsible sources
FSC™ C104740